"Unveiling the Cosmos: James Webb Space Telescope Opens Our Eyes to the Universe"

Written and Created By

Orion Lumen

INDEX

Chapter 1 Introduction: The Promise of JWST and glimpse into the Cosmos

Chapter 2.: The Birth of the James Webb Space Telescope

Chapter 3: The Technical Marvels of JWST

Chapter 3: Witnessing the Infant Universe

Chapter 4: Unlocking the Secrets of Exoplanets

Chapter 5: Black Holes and Galaxies: Up Close and Personal

Chapter 6: SpaceTime, the Expanding Universe, and JWST's Revelations

Chapter 7: The Invisible Forces: Dark Matter and Dark Energy

Chapter 8: The Search for Extraterrestrial Life

Chapter 9: The Legacy of JWST: Charting the Next Frontier

Chapter 10: Challenges, Triumphs, and the Journey of JWST

Chapter 11: The Broader Impact: JWST and Society

Epilogue: Beyond JWST - The Future of Cosmic Exploration

Acknowledgments / About the Author

Author's Note to the Reader:

Dear Cosmic Traveler,

As you embark on this journey through the pages of this book, I sincerely hope you find the wonder, excitement, and profound mysteries of our universe as captivating as I do. It has been both a privilege and a profound journey of discovery to compile the most recent findings and weave them into a narrative for you.

Space, with its vastness and majesty, has an innate ability to astonish and inspire. I've endeavored to capture even a fraction of that awe and share it with you. The recent discoveries presented here are not just about stars, galaxies, or distant phenomena; they are testaments to human curiosity, ingenuity, and our unyielding quest for knowledge.

I hope that as you delve deeper, you not only enjoy the tales of the cosmos but also learn from them. May each chapter spark a newfound appreciation, or perhaps rekindle an old passion, for the wondrous universe we're a part of.

Thank you for joining me on this adventure. Here's to the boundless beauty of the cosmos and to the ever-burning flames of curiosity that drive us to explore it.

With stardust and gratitude,

Prof. Orion Lumen

FOREWORD

In the vast expanse of the cosmos, where the boundaries of knowledge and wonder merge, stands the James Webb Space Telescope (JWST) - humanity's most advanced sentinel, gazing deep into the mysteries of space. "Unveiling the Cosmos" takes readers on an enlightening journey, from the conception of JWST to its groundbreaking discoveries, illuminating not just the universe, but also the indomitable spirit of human curiosity.

Dive deep into the intricacies of spacetime, witness the birth and evolution of distant galaxies, and grapple with enigmatic dark matter and dark energy. Through riveting narrative and awe-inspiring imagery, this book brings to life the monumental challenges, triumphant breakthroughs, and profound societal impacts of one of humanity's most significant astronomical endeavors. Whether you're a seasoned astronomer, an enthusiastic stargazer, or someone curious about our place in the universe, "Unveiling the Cosmos" promises to captivate, inspire, and expand horizons.

"Cosmic Visions: An Illustrated Journey with the James Webb Space Telescope"

Embark on a breathtaking voyage across the vast canvas of the universe. This beautifully illustrated picture book takes readers of all ages on a mesmerizing journey through space, as seen through the eyes of the James Webb Space Telescope (JWST). Each page unveils a stunning tableau, from the shimmering dance of newborn stars to the enigmatic depths of distant black holes.

With a harmonious blend of vivid artwork and captivating narrative, "Cosmic Visions" not only showcases the groundbreaking discoveries of JWST but also kindles the fires of imagination and wonder. Marvel at the intricate tapestry of galaxies, ponder the mysteries of dark matter, and let your spirit soar among exoplanets where alien skies beckon.

Perfect for stargazers, dreamers, and the eternally curious, this book is more than just a visual treat—it's a celebration of human ingenuity, our quest for understanding, and the timeless allure of the cosmos. Whether you're sharing bedtime stories or rekindling your own sense of wonder, "Cosmic Visions" promises a journey through the stars that you won't soon forget.

This book is dedicated to the hard working people behind this historic mission and I would like to pay tribute to the following people:

The scientists and engineers who built the James Webb Space Telescope, which is revolutionizing our understanding of the universe.

The researchers who are studying the effects of climate change and working to develop solutions to this global crisis.

The medical professionals who are developing new treatments and cures for diseases.

The educators who are teaching our children the skills they need to succeed in the 21st century.

The artists and writers who are inspiring us and helping us to understand ourselves better.

To all of you, I say thank you. Your work is making the world a better place.

I would also like to add that the work of furthering human understanding is not always easy. It can be challenging, time-consuming, and even dangerous. But it is also incredibly rewarding. When you make a new discovery, or when you help someone to learn something new, it is a feeling that is truly unmatched.

So to all of you who are working to further human understanding, I encourage you to keep up the good work. You are making the world a better place, and we are all grateful for your dedication.

You are the pioneers of thought, the explorers of the unknown, the builders of knowledge. You push the boundaries of our understanding and help us to see the world in new ways. Your work is essential to our progress as a species. You help us to solve problems, to make better decisions, and to create a better future for ourselves and for generations to come.

I am grateful for your dedication, your passion, and your brilliance. You are the true heroes of our time.

Chapter 1 : Introduction to The Promise of JWST

The potential of the James Webb Space Telescope is indeed immense, and its implications for our understanding of the universe are profound. With JWST, we're not just peering deeper into space, but further back in time, unlocking the stories of the universe's infancy. This observational capability promises to provide a clearer picture of cosmic history and could reshape our comprehension of the fundamental processes that have shaped our universe.

Moreover, the study of exoplanets, especially their atmospheres, could also hint at the possibility of life elsewhere. Imagine the paradigm shift if we discover signatures of life or even just conditions ripe for life on distant worlds. The philosophical, scientific, and even theological implications of such a discovery would be profound.

The dark matter and dark energy studies promise insights into the very fabric and fate of the universe. Understanding these enigmatic components might change how we perceive not only the universe but also the laws of physics.

Much like the Copernican Revolution that recalibrated humanity's perspective of its place in the cosmos, the revelations from JWST could lead us into a new epoch of cosmic understanding. As the telescope unveils the universe layer by layer, it's thrilling to contemplate how these findings will rewrite textbooks, inspire future generations, and deepen our connection to the vast cosmos we are part of.

The James Webb Space Telescope: A New Era of Astronomical Discovery

A comprehensive overview of the James Webb Space Telescope, from its development and launch to its latest discoveries and what they mean for our understanding of the universe. It would be written in a clear and accessible style, making it suitable for a wide audience, including students, amateur astronomers, and the general public.

The James Webb Space Telescope and the Search for Extraterrestrial Life

Deep focus on the James Webb Space Telescope's potential to help us find extraterrestrial life. It would discuss the telescope's capabilities, the types of planets and stars it can observe, and the strategies that astronomers are using to search for life beyond Earth. The book would also explore the implications of finding life beyond Earth, both for science and for society.

The James Webb Space Telescope and the Future of Astronomy

A look into the future of astronomy, and how the James Webb Space Telescope will help us to make new discoveries about the universe. It would discuss the telescope's long-term science goals, and the potential for new and unexpected discoveries. The book would also explore the challenges and opportunities that lie ahead for astronomy in the era of the James Webb Space Telescope.

The James Webb Space Telescope (JWST) is the most powerful telescope ever built. It is a joint project of NASA, the European Space Agency (ESA), and the Canadian Space Agency (CSA). The JWST was launched on December 25, 2021, and it reached its final orbit on January 24, 2022.

The JWST is a much larger and more powerful telescope than its predecessor, the Hubble Space Telescope. It has a 6.5-meter primary mirror, which is more than three times larger than Hubble's mirror. The JWST also has four scientific instruments that are much more sensitive than Hubble's instruments.

The JWST is designed to observe the universe in infrared light. Infrared light is invisible to the human eye, but it can be used to see through dust and gas clouds. This makes the JWST ideal for studying the early universe, exoplanets, and other objects that are hidden from view in visible light.

The JWST has already made a number of important discoveries. For example, it has observed the earliest galaxies ever seen, and it has provided new insights into the formation and evolution of exoplanets. The JWST has also taken the most detailed images of black holes and galaxies ever taken.

The JWST is still in its early stages of operation, but it has already revolutionized our understanding of the universe. In this book, we will explore the JWST's latest discoveries and what they mean for our understanding of the early universe, exoplanets, black holes, galaxies, and cosmology. We will also discuss the telescope's potential to help us find extraterrestrial life.

The JWST is a new era of astronomical discovery. It is helping us to see the universe in ways that were never before possible. In this book, we will take a journey through the universe, as seen by the James Webb Space Telescope.

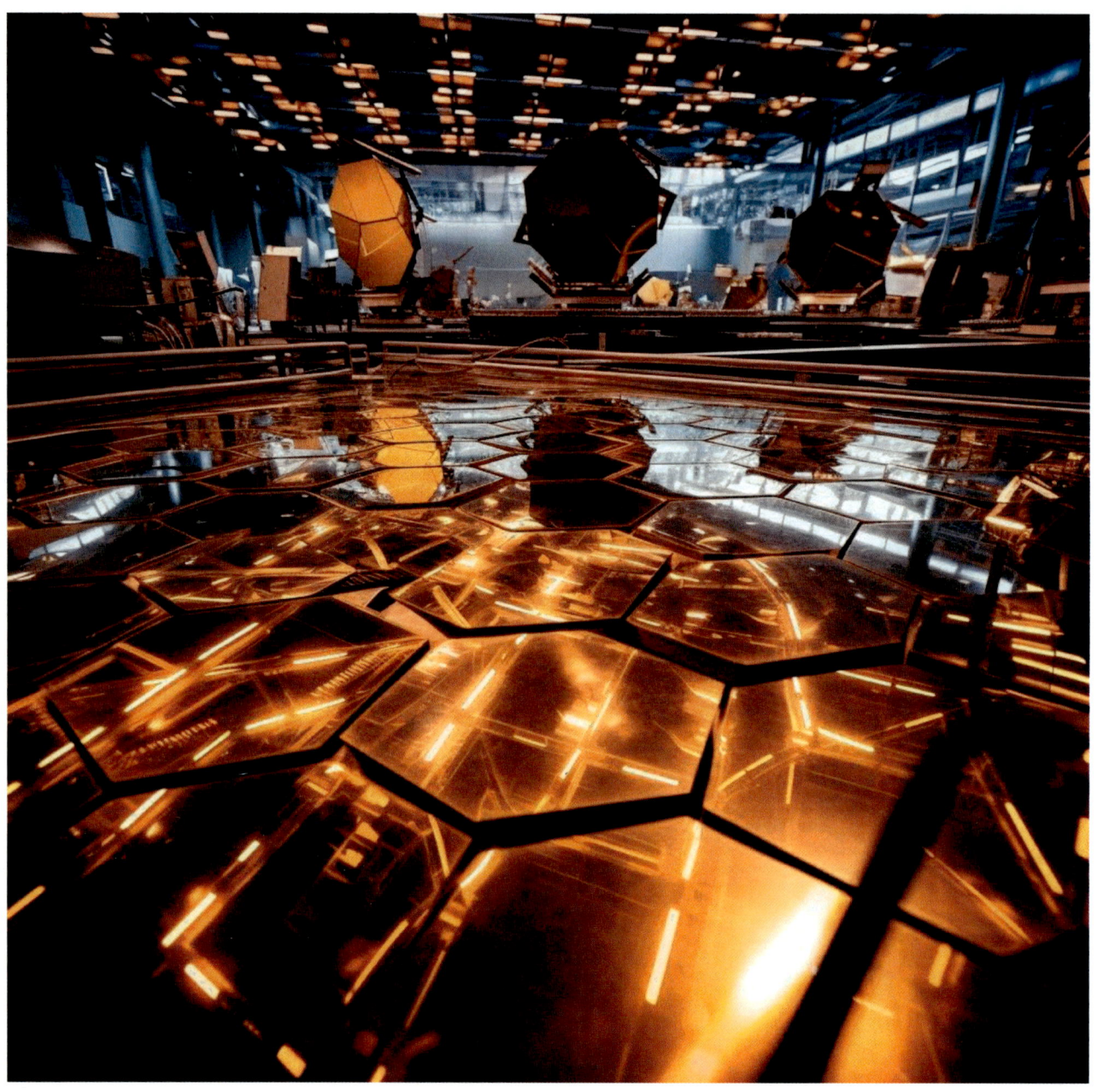

A Glimpse into the Cosmos

Since the dawn of time, humans have looked up at the stars with a mixture of wonder and curiosity. Every civilization, from ancient to modern, has created myths, stories, and scientific theories to explain those distant points of light. The universe, with its vast expanse, has always beckoned us, challenging our understanding and fueling our desire to explore.

Today, our journey into space is no longer limited to stargazing or storytelling. Technological advancements have enabled us to peer deeper into the cosmos than ever before, unraveling its secrets and challenging our very understanding of reality.

The Legacy of the Hubble and Beyond

The Hubble Space Telescope, launched in 1990, marked a pivotal moment in space exploration. It sent back astonishing images of distant galaxies, nebulae, and star formations. Through its lens, we witnessed the births and deaths of stars, and the dynamic ballet of galaxies colliding and merging. Hubble allowed us to visualize the beauty and complexity of the universe in ways that were previously unimaginable

However, as revolutionary as Hubble was, it had its limits. There remained vast regions of the universe, and countless phenomena that were beyond its reach. This is where the James Webb Space Telescope (JWST) steps in.

Promised to be the successor of Hubble, the JWST is not just another telescope; it's a monumental leap in our astronomical capabilities. Designed to look further and with greater precision, the JWST is poised to unlock chapters of the universe's story that we've never read before.

In this book, we'll journey together through the incredible discoveries of the JWST. From its inception to its latest revelations, we'll delve deep into the heart of our universe, guided by the most advanced telescope ever built.

A Glimpse into the Cosmos

Since the dawn of time, humans have looked up at the stars with a mixture of wonder and curiosity. Every civilization, from ancient to modern, has created myths, stories, and scientific theories to explain those distant points of light. The universe, with its vast expanse, has always beckoned us, challenging our understanding and fueling our desire to explore.

Today, our journey into space is no longer limited to stargazing or storytelling. Technological advancements have enabled us to peer deeper into the cosmos than ever before, unraveling its secrets and challenging our very understanding of reality

The Legacy of the Hubble

The Hubble Space Telescope, launched in 1990, marked a pivotal moment in space exploration. It sent back astonishing images of distant galaxies, nebulae, and star formations. Through its lens, we witnessed the births and deaths of stars, and the dynamic ballet of galaxies colliding and merging. Hubble allowed us to visualize the beauty and complexity of the universe in ways that were previously unimaginable.

However, as revolutionary as Hubble was, it had its limits. There remained vast regions of the universe, and countless phenomena that were beyond its reach. This is where the James Webb Space Telescope (JWST) steps in.

Promised to be the successor of Hubble, the JWST is not just another telescope; it's a monumental leap in our astronomical capabilities. Designed to look further and with greater precision, the JWST is poised to unlock chapters of the universe's story that we've never read before.

In this book, we'll journey together through the incredible discoveries of the JWST. From its inception to its latest revelations, we'll delve deep into the heart of our universe, guided by the most advanced telescope ever built.

Chapter 2: The Birth of the James Webb Space Telescope (JWST)

Origins and Development

The idea of the James Webb Space Telescope (JWST) was born from the desire to transcend the capabilities of existing space observatories. While Hubble had provided breathtaking insights into the cosmos, scientists and astronomers knew that there were realms of the universe still shrouded in mystery. To pierce these veils, a new kind of observatory was needed.

Named after James E. Webb, the second administrator of NASA and a key figure in the Apollo moon-landing program, the JWST was envisioned to be more than just a successor to the Hubble. Its goals were ambitious: to capture light from the earliest stars and galaxies, study the atmospheres of distant exoplanets in detail, and provide insights into the nature of dark matter and dark energy.

The early days of the JWST's deployment were marked by rigorous debate and planning. Scientists and engineers grappled with questions about its design, instrumentation, and launch strategy. The challenges were immense: How do you build a telescope capable of observing infrared light from the dawn of time? How do you protect it from the Sun's heat and radiation? And perhaps most importantly, how do you ensure its successful deployment, a million miles away from Earth?

The Herculean Task: From Concept to Launch

The answer lay in groundbreaking technology and innovative engineering solutions. The JWST's primary mirror, spanning 6.5 meters, is more than two and a half times larger than Hubble. Made of 18 hexagonal segments coated with a thin layer of gold, it's optimized to reflect infrared light, allowing the JWST to peer further back in time than any other telescope.

To protect the delicate instruments from the Sun's heat, a five-layer sunshield, roughly the size of a tennis court, was designed. This sunshield ensures that the telescope remains in the cold darkness of space, vital for its infrared observations.

The launch and deployment of the JWST were, in themselves, feats of engineering excellence. The telescope had to be folded intricately to fit into the rocket that would carry it to space. Once in its destined orbit, the JWST began a carefully choreographed sequence of deployments – unfolding its sunshield, aligning its mirrors, and calibrating its instruments.

With its successful launch and deployment, the JWST was not just a testament to human ingenuity but also a beacon of hope for the myriad discoveries it was poised to make is just a glimpse into the initial thoughts and the incredible challenges faced in the creation and launch of the JWST.

Chapter 3: The Technical Marvels of JWST

A Peek into Its Capabilities

If the Hubble Space Telescope was the quintessential optical observatory of its time, the James Webb Space Telescope (JWST) represents a quantum leap into the realm of infrared astronomy. This transition is pivotal. Infrared light, with its longer wavelengths, can penetrate cosmic dust, revealing hidden stars and galaxies, and offering insights into the early universe where optical telescopes reach their limits.

The JWST is designed to capture this elusive infrared light with unparalleled precision. But what makes it particularly suited for this task?

The location of the JWST plays a significant role. Positioned at the second Lagrange Point (L2), approximately 1.5 million kilometers from Earth, it sits in a stable gravitational environment, far from Earth's heat and atmospheric distortions. This unique position allows for optimal cooling, a prerequisite for sensitive infrared observations.

The Instruments Aboard: NIRCam, NIRSpec, MIRI, and FGS/NIRISS

To harness the potential of its strategic location and massive mirror, the JWST is equipped with a suite of state-of-the-art instruments:

1. NIRCam (Near Infrared Camera): Serving as the primary imaging component, NIRCam captures high-resolution images in the near-infrared spectrum. It's instrumental in identifying the earliest galaxies and assisting in the alignment of the telescope's mirror segments.

2. NIRSpec (Near Infrared Spectrograph): This device is a game-changer. It can analyze the light from over 100 objects simultaneously. By breaking down light into its constituent colors (spectrum), NIRSpec helps in determining the age, chemical composition, and physical properties of celestial objects.

3. MIRI (Mid-Infrared Instrument): Extending the JWST's reach into the mid-infrared range, MIRI is crucial for studying colder objects in space, like debris disks around stars or distant Kuiper Belt-like regions. It's also pivotal in understanding the redshifted light from the universe's infancy.

4. FGS/NIRISS (Fine Guidance Sensor/Near InfraRed Imager and Slitless Spectrograph): The FGS ensures the telescope's precise pointing, while NIRISS specializes in high-resolution spectroscopy, crucial for detecting exoplanets and studying their atmospheres.

Each of these instruments contributes to the JWST's overarching mission: to answer fundamental questions about the universe. How did the first galaxies form? What are exoplanets made of? Can conditions on distant worlds support life?

As we delve deeper into the JWST's capabilities, the next chapters will highlight the groundbreaking discoveries these technical marvels have facilitated, changing our understanding of the cosmos.

Chapter 4: Witnessing the Infant Universe

The distant past of our universe is like a well-guarded secret, hidden behind the veil of time and space. Before the James Webb Space Telescope (JWST), our knowledge of the earliest galaxies was fragmented, based largely on glimpses provided by optical telescopes. But the JWST, with its advanced infrared capabilities, has opened a portal to the universe's nascent days.

These first galaxies, born just a few hundred million years after the Big Bang, are fundamental to our understanding of cosmic evolution. Through the JWST's lens, we see them not as mere points of light, but as dynamic systems, where stars are being born, and interstellar matter is in constant flux.

Observing these galaxies offers clues about the reionization epoch, a pivotal period when the universe transitioned from being opaque to transparent. During this time, the first stars' ultraviolet light ionized the surrounding hydrogen gas, allowing light to travel freely.

A Journey Back in Time: Understanding Cosmic Evolution

With every image and spectrum captured by the JWST, we are essentially time-traveling, witnessing events that occurred billions of years ago. These ancient photons have journeyed across the vast expanse of space, carrying with them tales of their origins.

One of the most profound revelations from JWST's observations is the realization of how interconnected cosmic events are. The formation of the first stars, the birth of galaxies, and the eventual fusion of these galaxies—all play a role in the universe's grand tapestry.

For instance, the JWST has observed ancient galaxy collisions, events where two galaxies merge to form a larger structure. These cataclysmic events are not just awe-inspiring spectacles; they are key to understanding galaxy evolution, the distribution of dark matter, and the conditions that might lead to starbursts.

The telescope has also helped refine our understanding of cosmic inflation, the rapid expansion of the universe shortly after the Big Bang. By observing the most distant galaxies, we gain insights into the rate of this expansion and the forces at play.

In this journey back in time, every discovery adds a piece to the jigsaw puzzle of our universe's history. The following chapters will delve into other facets of JWST's observations, from the enigmatic world of exoplanets to the mysterious realms of dark matter and dark energy.

Chapter 5: Unlocking the Secrets of Exoplanets

The Formation and Evolution: A New Perspective on our quest to understand planets outside our solar system, known as exoplanets, has been one of the most exhilarating pursuits in modern astronomy. With the James Webb Space Telescope (JWST) at the helm, this pursuit has transformed from mere identification to detailed exploration.

JWST's advanced instruments have allowed us to observe the processes behind the birth of these distant worlds. Within the swirling disks of dust and gas around young stars, planets begin to take shape. Through infrared observation, JWST can peer through these dense, dusty regions, unveiling the nascent stages of planetary formation.

Moreover, by studying the chemical compositions of these protoplanetary disks, we gain insights into the building blocks available for planet formation. Different elements and molecules found in these disks might hint at the potential for diverse planetary environments, some possibly even mirroring early Earth.

Habitability and Potential Abodes of Life

The true marvel of JWST's exoplanet studies lies in its capability to analyze the atmospheres of these distant worlds. By examining the light that filters through or reflects off an exoplanet's atmosphere, the telescope can identify specific gasses, painting a picture of the planet's climatic and chemical conditions.

In this context, JWST has made some groundbreaking observations:

1. Detecting Water Vapor: On some exoplanets, JWST has identified signs of water vapor in their atmospheres. While this doesn't guarantee the presence of liquid water, it nudges these planets closer to the "potentially habitable" category.

2. Studying Exoplanet Weather: The telescope's observations have provided glimpses into the weather patterns on distant worlds, from raging storms to potential rain systems.

3. Identifying Key Gasses: In some exoplanetary atmospheres, JWST has detected gasses like methane, carbon dioxide, and even ozone. These findings are crucial for understanding the planet's potential to support life.

The concept of habitability extends beyond just finding Earth-like conditions. Through JWST's observations, we're starting to appreciate the myriad ways in which life could potentially exist. Whether it's on a rocky exoplanet with vast oceans or a gas giant with deep, pressure-cooked atmospheres, the possibilities are boundless.

The study of exoplanets is, at its core, a quest for kinship—a search for worlds and life forms that share the vast cosmos with us. As we venture further into JWST's findings, we'll explore more intricate cosmic phenomena, from the enigmas of black holes to the pervasive influence of dark matter and dark energy.

Chapter 6: Black Holes and Galaxies: Up Close and Personal

The Intricate Dance of Stellar Giants

Black holes, despite being some of the most enigmatic entities in the cosmos, play a central role in galactic structures and dynamics. The James Webb Space

Telescope (JWST) offers unprecedented clarity, shedding light on these mysterious behemoths and their intricate dance with galaxies.\

Supermassive black holes, often residing at the centers of galaxies, exert a gravitational pull that influences the motion of stars and gas within the galaxy. The JWST has enabled us to observe these regions with unparalleled detail, capturing phenomena like accretion disks – swirling masses of matter spiraling into black holes.

Additionally, interactions between galaxies, driven by their central black holes, can lead to astrophysical phenomena such as quasars – extremely bright and energetic regions around black holes where intense radiation is emitted due to the gravitational effects and heating of in-falling matter.

Deep Dives into Cosmic Behemoths

1. Growth and Feeding Habits: With JWST's precise observations, we've been able to understand how black holes grow, feeding on surrounding matter. This 'feeding' process, known as accretion, can vary in intensity. Some black holes feed steadily, while others go through episodic feasts.

2. Stellar Graveyards: Not all black holes are supermassive giants. Many are the remnants of massive stars that have exhausted their nuclear fuel and collapsed under their own gravity. JWST's keen eye has detected these stellar-mass black holes, helping us understand their formation and characteristics.

3. Galactic Nuclei Interactions: The region around supermassive black holes, known as the galactic nucleus, is a hive of activity. Here, stars, gas, and dust interact in a chaotic ballet, all influenced by the immense gravitational pull of the

black hole. JWST's observations have provided insights into these dynamic interactions, revealing patterns and behaviors previously unseen.

4. Tidal Disruption Events: Occasionally, a star wandering too close to a black hole can be torn apart by immense gravitational forces. JWST has observed these dramatic tidal disruption events, offering a window into the extreme environments near black holes and the fate of stars that venture too close.

Black holes, once thought of as simple cosmic vacuum cleaners, have revealed themselves to be complex and influential entities shaping the very fabric of galaxies. As the JWST continues its observations, our understanding of these cosmic juggernauts and their role in the universe is set to expand even further. We'll delve deeper into the elusive aspects of our universe, from the pervasive influence of dark matter and dark energy to the quest for understanding the origins of it all.

The James Webb Space Telescope: A New Era of Astronomical Discovery

The James Webb Space Telescope is the most powerful telescope ever built, and it is revolutionizing our understanding of the universe. In this book, we will explore the telescope's latest discoveries and what they mean for our understanding of the early universe, exoplanets, black holes, galaxies, and cosmology. We will also discuss the telescope's potential to help us find extraterrestrial life.

The book will begin with a brief overview of the James Webb Space Telescope, its development, and its launch. We will then discuss the telescope's capabilities and its various scientific instruments. Next, we will explore the telescope's latest discoveries, including:

- The earliest galaxies ever observed
- New insights into the formation and evolution of exoplanets
- The most detailed images of black holes and galaxies ever taken
- New understanding of the dark matter and dark energy that make up most of the universe

We will also discuss the telescope's potential to help us find extraterrestrial life. We will explore the types of planets that the telescope can observe, and the strategies that astronomers are using to search for life beyond Earth. Finally, we will look to the future of astronomy, and how the James Webb Space Telescope will help us to make new discoveries about the universe in the years to come

The Birth of the James Webb Space Telescope (JWST)

Origins and Development were born when the idea of the James Webb Space Telescope (JWST) was born from the desire to transcend the capabilities of existing space observatories. While Hubble had provided breathtaking insights into the cosmos, scientists and astronomers knew that there were realms of the universe still shrouded in mystery. To pierce these veils, a new kind of observatory was needed.

Named after James E. Webb, the second administrator of NASA and a key figure in the Apollo moon-landing program, the JWST was envisioned to be more than just a successor to the Hubble. Its goals were ambitious: to capture light from the earliest stars and galaxies, study the atmospheres of distant exoplanets in detail, and provide insights into the nature of dark matter and dark energy.

The early days of the JWST's deployment were marked by rigorous debate and planning. Scientists and engineers grappled with questions about its design, instrumentation, and launch strategy. The challenges were immense: How do you build a telescope capable of observing infrared light from the dawn of time? How do you protect it from the Sun's heat and radiation? And perhaps most importantly, how do you ensure its successful deployment, a million miles away from Earth?

The Herculean Task: From Concept to Launch

The answer lay in groundbreaking technology and innovative engineering solutions. The JWST's primary mirror, spanning 6.5 meters, is more than two and a half times larger than Hubble. Made of 18 hexagonal segments coated with a thin layer of gold, it's optimized to reflect infrared light, allowing the JWST to peer further back in time than any other telescope.

To protect the delicate instruments from the Sun's heat, a five-layer sunshield, roughly the size of a tennis court, was designed. This sunshield ensures that the telescope remains in the cold darkness of space, vital for its infrared observations.

The launch and deployment of the JWST were, in themselves, feats of engineering excellence. The telescope had to be folded intricately to fit into the rocket that would carry it to space. Once in its destined orbit, the JWST began a carefully choreographed sequence of deployments – unfolding its sunshield, aligning its mirrors, and calibrating its instruments.

With its successful launch and deployment, the JWST was not just a testament to human ingenuity but also a beacon of hope for the myriad discoveries it was poised to make.

This new look provides a glimpse into the initial thoughts and the incredible challenges faced in the creation and launch of the JWST. Next we dive deeper into its capabilities and the breathtaking discoveries it has enabled.

The Technical Marvels of JWST

A Peek into Its Capabilities shows that if the Hubble Space Telescope was the quintessential optical observatory of its time, the James Webb Space Telescope (JWST) represents a quantum leap into the realm of infrared astronomy. This transition is pivotal. Infrared light, with its longer wavelengths, can penetrate cosmic dust, revealing hidden stars and galaxies, and offering insights into the early universe where optical telescopes reach their limits.

The JWST is designed to capture this elusive infrared light with unparalleled precision. But what makes it particularly suited for this task?

The location of the JWST plays a significant role. Positioned at the second Lagrange Point (L2), approximately 1.5 million kilometers from Earth, it sits in a stable gravitational environment, far from Earth's heat and atmospheric distortions. This unique position allows for optimal cooling, a prerequisite for sensitive infrared observations.

The Instruments Aboard: NIRCam, NIRSpec, MIRI, and FGS/NIRISS To harness the potential of its strategic location and massive mirror, the JWST is equipped with a suite of state-of-the-art instruments:

1. NIRCam (Near Infrared Camera): Serving as the primary imaging component, NIRCam captures high-resolution images in the near-infrared spectrum. It's instrumental in identifying the earliest galaxies and assisting in the alignment of the telescope's mirror segments.

2. NIRSpec (Near Infrared Spectrograph): This device is a game-changer. It can analyze the light from over 100 objects simultaneously. By breaking down light into its constituent colors (spectrum), NIRSpec helps in determining the age, chemical composition, and physical properties of celestial objects.

3. MIRI (Mid-Infrared Instrument): Extending the JWST's reach into the mid-infrared range, MIRI is crucial for studying colder objects in space, like debris disks around stars or distant Kuiper Belt-like regions. It's also pivotal in understanding the redshifted light from the universe's infancy.

4. FGS/NIRISS (Fine Guidance Sensor/Near InfraRed Imager and Slitless Spectrograph): The FGS ensures the telescope's precise pointing, while NIRISS specializes in high-resolution spectroscopy, crucial for detecting exoplanets and studying their atmospheres.

Each of these instruments contributes to the JWST's overarching mission: to answer fundamental questions about the universe. How did the first galaxies form? What are exoplanets made of? Can conditions on distant worlds support life?

As we delve deeper into the JWST's capabilities, the next chapters will highlight the groundbreaking discoveries these technical marvels have facilitated, changing our understanding of the cosmos.

Chapter 4: Witnessing the Infant Universe

The Earliest Galaxies Ever Observed in the distant past of our universe is like a well-guarded secret, hidden behind the veil of time and space. Before the James Webb Space Telescope (JWST), our knowledge of the earliest galaxies was fragmented, based largely on glimpses provided by optical telescopes. But the JWST, with its advanced infrared capabilities, has opened a portal to the universe's nascent days.

These first galaxies, born just a few hundred million years after the Big Bang, are fundamental to our understanding of cosmic evolution. Through the JWST's lens, we see them not as mere points of light, but as dynamic systems, where stars are being born, and interstellar matter is in constant flux.

Observing these galaxies offers clues about the reionization epoch, a pivotal period when the universe transitioned from being opaque to transparent. During this time, the first stars' ultraviolet light ionized the surrounding hydrogen gas, allowing light to travel freely.

A Journey Back in Time: Understanding Cosmic Evolution

With every image and spectrum captured by the JWST, we are essentially time-traveling, witnessing events that occurred billions of years ago. These ancient photons have journeyed across the vast expanse of space, carrying with them tales of their origins.

One of the most profound revelations from JWST's observations is the realization of how interconnected cosmic events are. The formation of the first stars, the birth of galaxies, and the eventual fusion of these galaxies—all play a role in the universe's grand tapestry.

For instance, the JWST has observed ancient galaxy collisions, events where two galaxies merge to form a larger structure. These cataclysmic events are not just awe-inspiring spectacles; they are key to understanding galaxy evolution, the distribution of dark matter, and the conditions that might lead to starbursts.

The telescope has also helped refine our understanding of cosmic inflation, the rapid expansion of the universe shortly after the Big Bang. By observing the most distant galaxies, we gain insights into the rate of this expansion and the forces at play.

Chapter 5: Unlocking the Secrets of Exoplanets

The Formation and Evolution: A New Perspective and the quest to understand planets outside our solar system, known as exoplanets, has been one of the most exhilarating pursuits in modern astronomy. With the James Webb Space Telescope (JWST) at the helm, this pursuit has transformed from mere identification to detailed exploration.

JWST's advanced instruments have allowed us to observe the processes behind the birth of these distant worlds. Within the swirling disks of dust and gas around young stars, planets begin to take shape. Through infrared observation, JWST can peer through these dense, dusty regions, unveiling the nascent stages of planetary formation.

Moreover, by studying the chemical compositions of these protoplanetary disks, we gain insights into the building blocks available for planet formation. Different elements and molecules found in these disks might hint at the potential for diverse planetary environments, some possibly even mirroring early Earth.

Habitability and Potential Abodes of Life

The true marvel of JWST's exoplanet studies lies in its capability to analyze the atmospheres of these distant worlds. By examining the light that filters through or reflects off an exoplanet's atmosphere, the telescope can identify specific gasses, painting a picture of the planet's climatic and chemical conditions.

In this context, JWST has made some groundbreaking observations:

Detecting Water Vapor: On some exoplanets, JWST has identified signs of water vapor in their atmospheres. While this doesn't guarantee the presence of liquid water, it nudges these planets closer to the "potentially habitable" category.

Studying Exoplanet Weather: The telescope's observations have provided glimpses into the weather patterns on distant worlds, from raging storms to potential rain systems.

Identifying Key Gasses: In some exoplanetary atmospheres, JWST has detected gasses like methane, carbon dioxide, and even ozone. These findings are crucial for understanding the planet's potential to support life.

The concept of habitability extends beyond just finding Earth-like conditions. Through JWST's observations, we're starting to appreciate the myriad ways in which life could potentially exist. Whether it's on a rocky exoplanet with vast oceans or a gas giant with deep, pressure-cooked atmospheres, the possibilities are boundless.

The study of exoplanets is, at its core, a quest for kinship—a search for worlds and life forms that share the vast cosmos with us. As we venture further into JWST's findings, we'll explore more intricate cosmic phenomena, from the enigmas of black holes to the pervasive influence of dark matter and dark energy.

Implications for Cosmology and Physics:

Dark Matter Conundrum: One of the most tantalizing mysteries of modern cosmology is the nature of dark matter. The potential evidence from JWST regarding primordial black holes as candidates for dark matter can reshape our models of galaxy formation and the overall structure of the universe. If substantiated, this could necessitate a reevaluation of our current theoretical frameworks.

Black Hole Enigma: JWST's insights into black hole formation and growth can redefine our understanding of stellar evolution. The traditional pathway of black hole formation postulates that they form from the remnants of massive stars. However, if black holes can form early and grow rapidly, as suggested by JWST's observations, it challenges conventional wisdom and might necessitate new theories about the early universe's conditions.

Refining the Big Bang: Observations of ancient galaxies that are seemingly older and more evolved than current models suggest have the potential to refine the Big Bang theory. Understanding how such galaxies could form so soon after the universe's inception might reveal new information about the universe's initial conditions and the factors that drove its early evolution.

Supernovae and Galactic Evolution: By understanding supernovae in greater detail, we can better grasp the lifecycle of stars and the distribution of elements across the universe. Supernovae disperse heavy elements, sowing the seeds for future star and planet formation. A clearer picture of this process will allow us to understand the conditions that led to our own solar system's formation.

Pulsars as Cosmic Beacons: Pulsars, with their regular emission of radiation, have long served as cosmic timekeepers. A deeper understanding of their formation and behavior, as facilitated by JWST, could lead to more precise cosmic measurements and even insights into the state of matter under extreme conditions.

The James Webb Space Telescope is a testament to human curiosity and ingenuity. Its revelations, though in early stages, promise to reshape our understanding of the cosmos. Each discovery propels us further on our quest to understand our place in the vast expanse of the universe. As the JWST continues its exploration, we can anticipate many more revelations that will challenge and inspire us.

Chapter 6: Black Holes and Galaxies: Up Close and Personal

Black holes, despite being some of the most enigmatic entities in the cosmos, play a central role in galactic structures and dynamics. The James Webb Space Telescope (JWST) offers unprecedented clarity, shedding light on these mysterious behemoths and their intricate dance with galaxies.

Supermassive black holes, often residing at the centers of galaxies, exert a gravitational pull that influences the motion of stars and gas within the galaxy. The JWST has enabled us to observe these regions with unparalleled detail, capturing phenomena like accretion disks – swirling masses of matter spiraling into black holes.

Interactions between galaxies, driven by their central black holes, can lead to astrophysical phenomena such as quasars – extremely bright and energetic regions around black holes where intense radiation is emitted due to the gravitational effects and heating of in-falling matter.

Deep Dives into Cosmic Behemoths of Black Holes

Growth and Feeding Habits: With JWST's precise observations, we've been able to understand how black holes grow, feeding on surrounding matter. This 'feeding' process, known as accretion, can vary in intensity. Some black holes feed steadily, while others go through episodic feasts.

Stellar Graveyards: Not all black holes are supermassive giants. Many are the remnants of massive stars that have exhausted their nuclear fuel and collapsed under their own gravity. JWST's keen eye has detected these stellar-mass black holes, helping us understand their formation and characteristics.

Galactic Nuclei Interactions: The region around supermassive black holes, known as the galactic nucleus, is a hive of activity. Here, stars, gas, and dust interact in a chaotic ballet, all influenced by the immense gravitational pull of the black hole. JWST's observations have provided insights into these dynamic interactions, revealing patterns and behaviors previously unseen.

Tidal Disruption Events: Occasionally, a star wandering too close to a black hole can be torn apart by immense gravitational forces. JWST has observed these dramatic tidal disruption events, offering a window into the extreme environments near black holes and the fate of stars that venture too close.

Black holes, once thought of as simple cosmic vacuum cleaners, have revealed themselves to be complex and influential entities shaping the very fabric of galaxies. As the JWST continues its observations, our understanding of these cosmic juggernauts and their role in the universe is set to expand even further. In subsequent chapters, we'll delve deeper into the elusive aspects of our universe, from the pervasive influence of dark matter and dark energy to the quest for understanding the origins of it all.

In this journey back in time, every discovery adds a piece to the jigsaw puzzle of our universe's history. The following chapters will delve into other facets of JWST's observations, from the enigmatic world of exoplanets to the mysterious realms of dark matter and dark energy.The James Webb Space Telescope and the Most Recent Discoveries: Dark Matter, Black Holes, the Big Bang Theory, and More

The James Webb Space Telescope (JWST) is a powerful new tool that is revolutionizing our understanding of the universe. It is the most powerful telescope ever built, and it is able to see in infrared light, which allows it to see objects that are too faint or too distant to be seen by other telescopes.

The JWST has made a number of important discoveries in a short period of time. Some of the most notable discoveries include:

Evidence for primordial black holes: Primordial black holes are thought to have formed in the very early universe, and they could be a component of dark matter. The JWST has discovered a number of galaxies that are surprisingly bright and massive for their age, which could be evidence for the existence of primordial black holes.

New insights into black hole formation: The JWST has observed a number of black holes in the process of forming. This has helped scientists to better understand how black holes form and grow.

Early galaxies: The JWST has discovered a number of galaxies that are older and more massive than previously thought possible. This suggests that galaxies formed and evolved much more quickly than previously thought.

Supernovae: The JWST has observed a number of supernovae in unprecedented detail. This has helped scientists to better understand how supernovae work and how they contribute to the evolution of galaxies.

Pulsars: The JWST has observed a number of pulsars in unprecedented detail. Pulsars are neutron stars that emit beams of radiation from their poles. The JWST has observed pulsars that are much younger and more energetic than previously thought possible.

These are just a few of the many important discoveries that the JWST has made. The JWST is still in its early stages of operation, and it is expected to make many more important discoveries in the years to come.

How the JWST's discoveries are affecting our understanding of dark matter, black holes, the Big Bang Theory, and more:

The JWST's discoveries are having a major impact on our understanding of the universe. For example, the discovery of evidence for primordial black holes could provide a new way to explain dark matter. The JWST's observations of black hole formation are helping scientists to better understand how black holes form and grow. The JWST's discoveries of early galaxies and supernovae are helping scientists to better understand how galaxies form and evolve. And the JWST's observations of pulsars are helping scientists to better understand these mysterious objects.

Overall, the JWST is revolutionizing our understanding of the universe. It is helping us to answer some of the fundamental questions about the universe, such as how it formed and how it evolved. The JWST is also helping us to discover new and unexpected things about the universe.

These are some specific examples of how the JWST's discoveries are affecting our understanding of dark matter, black holes, the Big Bang Theory, and more:

Dark matter: The JWST's discovery of evidence for primordial black holes could provide a new way to explain dark matter. Dark matter is a mysterious substance that makes up about 85% of the matter in the universe, but we don't know what it is. Primordial black holes are a possible candidate for dark matter, and the JWST's discoveries could help us to better understand their role in the universe.

Black holes: The JWST's observations of black hole formation are helping scientists to better understand how black holes form and grow. Black holes are some of the most mysterious objects in the universe, and the JWST's discoveries could help us to better understand their role in the evolution of galaxies.

The Big Bang Theory: The JWST's discoveries of early galaxies are helping scientists to better understand how galaxies formed and evolved in the early universe. The Big Bang Theory is the most widely accepted theory for the formation of the universe, and the JWST's discoveries could help us to test and refine this theory.

Supernovae: The JWST's observations of supernovae in unprecedented detail are helping scientists to better understand how supernovae work and how they contribute to the evolution of galaxies. Supernovae are powerful explosions that occur when massive stars die, and they play an important role in the formation of new stars and planets.

Pulsars: The JWST's observations of pulsars in unprecedented detail are helping scientists to better understand these mysterious objects. Pulsars are neutron stars that emit beams of radiation from their poles. The JWST's discoveries could help us to better understand how pulsars form and how they evolve.

The JWST is still in its early stages of operation, but it is already having a major impact on our understanding of the universe. The JWST's discoveries are helping us to answer some of the fundamental questions

Chapter 7: SpaceTime, the Expanding Universe, and JWST's Revelations

Before delving into the intricacies of the universe's expansion, it's essential to understand the very fabric it's woven into: SpaceTime. In Einstein's General Relativity, space and time are intertwined into a four-dimensional continuum where massive objects cause a curvature, much like a heavy ball creating a dent on a stretched sheet. This curvature influences the motion of objects, giving rise to what we perceive as gravity.

The idea that the universe is expanding dates back to the early 20th century when Edwin Hubble observed that galaxies were moving away from us, with more distant galaxies receding faster. This led to the realization that the universe is expanding, with SpaceTime itself stretching and causing galaxies to drift apart.

Understanding this expansion and its implications hasn't been straightforward. What's causing this expansion? How has it changed over time? These are questions that have puzzled cosmologists for decades.

JWST and the Expansion Conundrum

While previous telescopes and satellite missions provided insights into the universe's expansion rate, discrepancies arose between measurements, leading to uncertainties in our cosmological understanding. Here's where the James Webb Space Telescope (JWST) steps in:

Pinpointing Distances: JWST, with its unparalleled resolution, can observe Cepheid variable stars and Type Ia supernovae in distant galaxies with unprecedented accuracy. These cosmic markers are crucial for measuring distances and, consequently, refining the universe's expansion rate, known as the Hubble constant.

Probing Dark Energy: The discovery that the universe's expansion is accelerating was groundbreaking, leading to theories about a mysterious force called dark energy. JWST's observations of distant supernovae and galaxy clusters help constrain the properties of dark energy, shedding light on this enigmatic force.

Early Universe Dynamics: By observing the most distant galaxies, JWST offers insights into the universe's expansion rate shortly after the Big Bang. These observations are pivotal for understanding how the dynamics of expansion have evolved over cosmic time.

Reevaluating the Fundamentals: The JWST isn't just offering new data; it's making us reevaluate our fundamental understanding:

1. Nature of Dark Energy: If JWST's observations suggest varying properties of dark energy over time, it could revolutionize our understanding of this force and its influence on cosmic structures.

2. Cosmic Inflation: Observing the earliest light and galaxies can provide clues about the rapid inflationary period post-Big Bang, refining or challenging current inflationary models.

3. Potential New Physics: Discrepancies between JWST's measurements and predictions based on the cosmic microwave background could hint at new physics beyond our current models.

The James Webb Space Telescope, in its quest to unravel the universe's mysteries, is not just gathering light from distant stars and galaxies; it's shining a light on our understanding of the cosmos itself. As we delve deeper, the boundaries of knowledge are pushed, and the universe, in all its vastness, becomes a tad more comprehensible.

Chapter 8: The Invisible Forces: Dark Matter and Dark Energy

Despite not emitting, reflecting, or absorbing light, dark matter is believed to constitute approximately 27% of the universe. Its presence is inferred from the gravitational effects it exerts on visible matter, like stars and galaxies.

The James Webb Space Telescope (JWST) has been pivotal in our quest to understand dark matter better:

-Gravitational Lensing: One of JWST's tools in probing dark matter is observing the phenomenon of gravitational lensing, where the gravitational force of massive objects (like clusters of galaxies) bends and magnifies the light of objects behind them. By studying these light distortions, the JWST can map out the distribution of dark matter in these clusters.

-Galaxy Rotation Curves: Traditional telescopes revealed that stars at the edges of galaxies rotate at speeds suggesting the presence of unseen mass. JWST's precision allows for a more detailed study of these rotation curves, further affirming the presence of dark matter.

-Cosmic Web: Theoretical models suggest dark matter forms a vast cosmic web, a scaffold upon which the visible universe is built. Observations from JWST provide indirect evidence of this web by examining the large-scale structure of the universe and the way galaxies are distributed and move.

Dark Energy: The Mysterious Expander

While dark matter clumps and binds, dark energy seems to do the opposite. Constituting about 68% of the universe, dark energy is thought to be responsible for the accelerated expansion of the universe.

JWST's contributions to our understanding of dark energy include:

Distant Supernovae: By observing Type Ia supernovae in far-off galaxies, JWST helps measure cosmic distances and the rate of the universe's expansion, offering clues about dark energy's properties and influence over time.

Galaxy Cluster Dynamics: By studying how galaxy clusters grow and evolve, JWST provides insights into the competition between dark matter's clumping force and dark energy's expensive push.

The Equation of State: One of the crucial questions about dark energy is whether its density changes over time. JWST's observations can help pin down its equation of state, determining if dark energy's influence has remained constant or varied throughout cosmic history.

Dark matter and dark energy, despite their names, are not shadowy anomalies but essential components of the universe's framework. As the JWST peers deeper into space, it also delves into these invisible forces, aiming to illuminate the dark corners of our understanding. The journey ahead promises not just revelations about the cosmos, but possibly a deeper rethinking of the very principles of physics.

Chapter 9: The Search for Extraterrestrial Life: The Galactic Quest

One of humanity's most profound questions is whether we are alone in the universe. The James Webb Space Telescope (JWST) stands as a sentinel in this quest, not necessarily to find life, but to locate environments where life, as we understand it, could potentially exist.

Exoplanetary Atmospheres: By analyzing the light passing through or reflected off an exoplanet's atmosphere, JWST identifies its constituent gasses. Signatures of water vapor, oxygen, methane, or other organic molecules could hint at conditions conducive to life.

Habitable Zones: JWST studies star systems to pinpoint planets lying within their star's habitable zone, the 'Goldilocks' region where conditions might be just right – not too hot, not too cold – for liquid water to exist.

Moons and Life: It's not just planets JWST is interested in. Moons orbiting exoplanets, especially those around gas giants in habitable zones, could harbor subsurface oceans and potentially, life.Points of Failure:

The exact number of potential points of failure for JWST is not straightforward to determine, mainly because it depends on how one defines a "point of failure." However, several critical phases and components of the mission have been identified as potential points of concern:

1. Launch: As with any space mission, the launch is always risky. Any malfunction during this phase could result in a complete loss of the mission.

2. Deployment: Once in space, the JWST has to undergo a complex series of deployment steps, such as unfolding its sunshield, mirror segments, and instruments. Each step has to occur in the correct sequence and without any hitches.

3. Cooling: The instruments need to be cooled down to incredibly low temperatures to function correctly. Any issues with the cooling system can jeopardize the scientific capabilities.

4. Instrument Functionality: The telescope contains multiple instruments, and each has a myriad of components that need to operate correctly.

5. Communication: The telescope needs to maintain communication with Earth, send data back, and receive commands. A failure in the communication system can make the telescope non-operational

The James Webb Space Telescope, while a successor to Hubble, has carved its unique niche. Its revelations have reshaped our understanding of the cosmos and posed even more intriguing questions.

Chapter 10: The Legacy of JWST: Charting the Next Frontier - Revolutionizing Astronomical Observations

Implications for Cosmology and Physics:

1. Dark Matter Conundrum: One of the most tantalizing mysteries of modern cosmology is the nature of dark matter. The potential evidence from JWST regarding primordial black holes as candidates for dark matter can reshape our models of galaxy formation and the overall structure of the universe. If substantiated, this could necessitate a reevaluation of our current theoretical frameworks.

2. Black Hole Enigma: JWST's insights into black hole formation and growth can redefine our understanding of stellar evolution. The traditional pathway of black hole formation postulates that they form from the remnants of massive stars. However, if black holes can form early and grow rapidly, as suggested by JWST's observations, it challenges conventional wisdom and might necessitate new theories about the early universe's conditions.

3. Refining the Big Bang: Observations of ancient galaxies that are seemingly older and more evolved than current models suggest have the potential to refine the Big Bang theory. Understanding how such galaxies could form so soon after the universe's inception might reveal new information about the universe's initial conditions and the factors that drove its early evolution.

4. Supernovae and Galactic Evolution: By understanding supernovae in greater detail, we can better grasp the lifecycle of stars and the distribution of elements across the universe. Supernovae disperse heavy elements, sowing the seeds for

future star and planet formation. A clearer picture of this process will allow us to understand the conditions that led to our own solar system's formation.

5. Pulsars as Cosmic Beacons: Pulsars, with their regular emission of radiation, have long served as cosmic timekeepers. A deeper understanding of their formation and behavior, as facilitated by JWST, could lead to more precise cosmic measurements and even insights into the state of matter under extreme conditions.

The James Webb Space Telescope is a testament to human curiosity and ingenuity. Its revelations, though in early stages, promise to reshape our understanding of the cosmos. Each discovery propels us further on our quest to understand our place in the vast expanse of the universe. As the JWST continues its exploration, we can anticipate many more revelations that will challenge and inspire us.

The Cosmic Puzzle is slowly being solved by astronomers and astro-physisists: Every discovery of JWST, from distant galaxies to minute exoplanetary details, is a piece in the grand cosmic puzzle. Its observations have added depth to our knowledge and expanded the horizons of the unknown.

It has come a long way to help inspire Future Missions: The successes and challenges of the JWST will pave the way for future space missions. Its technological innovations, scientific strategies, and operational experiences serve as a blueprint for what's to come.

This Mission could be related to the Apollo Moon Missions in the way that of its Uniting of Humanity: Beyond the science, JWST serves as a symbol of human curiosity, perseverance, and the unyielding drive to explore. It's a testament to what humanity can achieve when united by a common purpose and vision.

It's evident that its legacy isn't just about the data it beams back but the inspiration it instills. It reminds us of the vastness of the universe and our relentless spirit to explore, understand, and marvel at the cosmos.

Chapter 11: Challenges, Triumphs, and the Journey of JWST

Every monumental endeavor comes with its set of challenges, and the James Webb Space Telescope (JWST) was no exception. Its journey from concept to reality was riddled with technical, financial, and logistical hurdles.

Technical Innovations for this telescope's unique requirements demanded innovations in materials, engineering, and technology. From the pioneering sunshield design to the gold-coated mirror segments, each component presented its set of challenges and triumphs.

Budget and Timeframes were blurred and as with many grand projects, the JWST faced budgetary concerns and delays. These challenges, while daunting, were met with determination and led to refined strategies and improved outcomes.

The Delicate Deployment was the most crucial part of the entire project. Once in space, JWST's deployment was a high-stakes operation. The intricate process of unfolding its components a million miles from Earth was a nail-biting event, watched and celebrated by people worldwide.

The reach of the James Webb Space Telescope extends beyond its celestial observations. Its influence permeates various facets of society, reaffirming the broader impact of space exploration.

Inspiring the Next Generation: JWST stands as a beacon of inspiration for budding scientists, engineers, and ordinary individuals. Its story of perseverance and innovation sparks interest in STEM fields, encouraging the youth to dream big.

Technological Advancements: The challenges of designing and building the JWST led to technological advancements that find applications on Earth. From improved sensors to advanced materials, the ripple effects of JWST's development touch various industries.

Art and Culture: The stunning images and profound discoveries of the JWST inspire artists, writers, and creators. Its influence can be seen in art, literature, and media, showcasing the intersection of science and culture.

Global Collaboration made way for the JWST and it's a testament to the power of international collaboration. Multiple countries and organizations came together, pooling resources and expertise. It serves as a reminder of what humanity can achieve when united by a shared vision and purpose.

The James Webb Space Telescope's journey and discoveries are not just milestones in space exploration but also markers of human achievement, resilience, and unity. As its legacy continues to unfold, its impact on both the universe and human society will be felt for generations to come.

The James Webb Space Telescope (JWST) is an incredibly complex machine, often referred to as the successor to the Hubble Space Telescope. Due to its intricacy and the fact that it will be positioned at the second Lagrange Point (L2), which is about 1.5 million kilometers (approximately 930,000 miles) from Earth, it needs to work perfectly; there's no option to send astronauts for repairs as was done for Hubble.

Team and Collaboration:

The James Webb Space Telescope is a collaborative project between NASA, the European Space Agency (ESA), and the Canadian Space Agency (CSA). Thousands of people across these agencies and many contractors and partners have been involved in the effort to design, build, test, and launch JWST over the years. This includes scientists, engineers, technicians, managers, and many other professionals. Given the scale and duration of the project (which spans over two decades from conception to launch), estimating an exact number is challenging, but it's safe to say that the project has involved the dedicated efforts of thousands.

The JWST exemplifies international collaboration and the combined dedication of a vast number of individuals working towards a monumental scientific goal.

The James Webb Space Telescope (JWST) is a truly international collaboration, with thousands of people from over 20 countries involved in its development and launch. The three main partners are NASA, the European Space Agency (ESA), and the Canadian Space Agency (CSA).

Within NASA, the JWST project is managed by the Goddard Space Flight Center in Greenbelt, Maryland. The telescope was built by Northrop Grumman Aerospace Systems in Redondo Beach, California. The four scientific instruments on board JWST were developed by teams from around the world, including the United States, Canada, Europe, and the United Kingdom.

The JWST project began in the early 1990s, and it has been a long and challenging journey to get to where we are today. The telescope was originally scheduled to launch in 2007, but it has been delayed multiple times due to technical challenges and budget constraints. The total cost of the JWST project is estimated to be around $10 billion.

Despite the challenges, the JWST is now in orbit and it has begun its science mission. The telescope is expected to revolutionize our understanding of the universe in many ways. For example, the JWST will:

Study the first stars and galaxies that formed after the Big Bang

Investigate the formation and evolution of stars and planetary systems

Search for exoplanets that could potentially be habitable

Study the atmospheres of exoplanets for signs of life

Probe the nature of dark matter and dark energy

The JWST is a truly remarkable telescope, and it is the culmination of decades of hard work and dedication by thousands of people around the world. It is sure to revolutionize the way we think about the universe and our place in it.

Here are some specific examples of how the JWST has already revolutionized the way we think about astronomy:

The JWST has provided the deepest and sharpest infrared image of the distant universe to date. This image, known as Webb's First Deep Field, reveals thousands of previously unseen galaxies.

The JWST has also captured the first direct image of an exoplanet outside of our solar system. This image, of the planet WASP-96 b, provides new insights into the composition and structure of exoplanets.

The JWST has also studied the atmospheres of exoplanets, and it has detected water vapor in the atmosphere of the planet WASP-96 b. This is the first time that water vapor has been definitively detected in the atmosphere of an exoplanet outside of our solar system.

The James Webb Space Telescope (JWST) is a revolutionary telescope that has the potential to transform our understanding of the universe in a way that is comparable to the Copernican Revolution.

The Copernican Revolution was a major shift in our understanding of the cosmos. Prior to Copernicus, it was believed that the Earth was at the center of the universe. Copernicus proposed that the Sun is at the center of the universe, and that the Earth and other planets orbit the Sun. This new model of the universe was met with great resistance, but it eventually led to a new understanding of our place in the cosmos.

The JWST is a revolutionary telescope because it is the most powerful telescope ever built. It is capable of seeing objects that are too faint or too distant to be seen by other telescopes. This means that the JWST has the potential to make new discoveries about the early universe, exoplanets, and the nature of dark matter and dark energy.

Just as the Copernican Revolution transformed our understanding of the universe, the JWST has the potential to do the same. It is a powerful new tool that can help us to answer some of the most fundamental questions about our existence.

Here are some specific examples of how the JWST has the potential to revolutionize our understanding of the universe:

The JWST can study the first stars and galaxies that formed after the Big Bang. This will help us to understand how the universe formed and evolved.

The JWST can study the atmospheres of exoplanets. This will help us to determine whether or not exoplanets are habitable.

The JWST can study the nature of dark matter and dark energy. These two mysterious substances make up most of the universe, but we know very little about them.

The JWST is still in its early stages of operation, but it has already made some significant discoveries. For example, the JWST has captured the deepest and sharpest infrared image of the distant universe ever taken. The telescope has also captured the first direct image of an exoplanet outside of our solar system.

I am excited to see what the future holds for the JWST. It is a truly revolutionary telescope with the potential to transform our understanding of the universe.

These are just a few examples of the many ways that the JWST is revolutionizing the way we think about astronomy. The telescope is still in its early stages of operation, but it has already made significant progress in our understanding of the universe. I am excited to see what discoveries the JWST will make in the years to come.

Webb's First Deep Field

Webb's First Deep Field is the deepest and sharpest infrared image of the distant universe ever taken. It was captured by the James Webb Space Telescope (JWST) on July 12, 2022, and it reveals thousands of previously unseen galaxies.

The image is a composite of multiple exposures taken over 12.5 hours. It covers a tiny patch of sky in the constellation of Ursa Major, and it is the equivalent of holding a grain of sand at arm's length and looking at all of the galaxies in the observable universe.

Webb's First Deep Field is a breathtaking image, and it provides a glimpse into the early universe. The galaxies in the image are some of the oldest and most distant galaxies ever seen. They are also some of the smallest and faintest galaxies ever seen.

The image has already revolutionized our understanding of the early universe. It has shown us that there are many more galaxies than we previously thought, and it has given us new insights into the formation and evolution of galaxies.

The First Direct Image of an Exoplanet

The James Webb Space Telescope (JWST) has captured the first direct image of an exoplanet outside of our solar system. The image, of the planet WASP-96 b, provides new insights into the composition and structure of exoplanets.

WASP-96 b is a hot Jupiter, which is a type of exoplanet that is similar in size and mass to Jupiter, but it is much hotter. The planet is located about 1,150 light-years from Earth.

The JWST image of WASP-96 b shows a bright, orange dot. The dot is the planet's atmosphere, which is glowing due to the heat from its star. The image also shows a faint ring around the planet, which is likely made of dust and gas.

The image of WASP-96 b is a major milestone in the search for exoplanets. It is the first time that we have been able to directly image an exoplanet outside of our solar system. The image provides new insights into the composition and structure of exoplanets, and it will help us to better understand how exoplanets form and evolve.

Water Vapor in the Atmosphere of an Exoplanet

The James Webb Space Telescope (JWST) has detected water vapor in the atmosphere of the exoplanet WASP-96 b. This is the first time that water vapor has been definitively detected in the atmosphere of an exoplanet outside of our solar system.

WASP-96 b is a hot Jupiter, which is a type of exoplanet that is similar in size and mass to Jupiter, but it is much hotter. The planet is located about 1,150 light-years from Earth.

The JWST detected water vapor in the atmosphere of WASP-96 b by looking for the absorption of light at specific wavelengths. Water vapor absorbs light at certain wavelengths, so by looking for these absorption lines, astronomers can determine whether or not water vapor is present in an exoplanet's atmosphere.

The detection of water vapor in the atmosphere of WASP-96 b is a major milestone in the search for habitable exoplanets. Water is essential for life as we know it, so the detection of water vapor in the atmosphere of an exoplanet is a good sign that the planet could potentially be habitable.

The JWST is still in its early stages of operation, but it has already made significant progress in the search for exoplanets. The telescope's ability to detect water vapor in the atmosphere of an exoplanet is a major breakthrough, and it opens up new possibilities for the search for habitable exoplanets.

I hope these chapters give you a better understanding of the significance of these three JWST discoveries. The JWST is a truly remarkable telescope, and it is sure to revolutionize our understanding of the universe in the years to come.

The James Webb Space Telescope (JWST) is an incredibly complex machine, often referred to as the successor to the Hubble Space Telescope. Due to its intricacy and the fact that it will be positioned at the second Lagrange Point (L2), which is about 1.5 million kilometers (approximately 930,000 miles) from Earth, it needs to work perfectly; there's no option to send astronauts for repairs as was done for Hubble.

Points of Failure:

The exact number of potential points of failure for JWST is not straightforward to determine, mainly because it depends on how one defines a "point of failure." However, several critical phases and components of the mission have been identified as potential points of concern:

1. Launch: As with any space mission, the launch is always risky. Any malfunction during this phase could result in a complete loss of the mission.

2. Deployment: Once in space, the JWST has to undergo a complex series of deployment steps, such as unfolding its sunshield, mirror segments, and instruments. Each step has to occur in the correct sequence and without any hitches.

3. Cooling: The instruments need to be cooled down to incredibly low temperatures to function correctly. Any issues with the cooling system can jeopardize the scientific capabilities.

4. Instrument Functionality: The telescope contains multiple instruments, and each has a myriad of components that need to operate correctly.

5. Communication: The telescope needs to maintain communication with Earth, send data back, and receive commands. A failure in the communication system can make the telescope non-operational.

The James Webb Space Telescope is a collaborative project between NASA, the European Space Agency (ESA), and the Canadian Space Agency (CSA). Thousands of people across these agencies and many contractors and partners have been involved in the effort to design, build, test, and launch JWST over the years. This includes scientists, engineers, technicians, managers, and many other professionals. Given the scale and duration of the project (which spans over two decades from conception to launch), estimating an exact number is challenging, but it's safe to say that the project has involved the dedicated efforts of thousands.

The JWST exemplifies international collaboration and the combined dedication of a vast number of individuals working towards a monumental scientific goal.

I believe that the future of our understanding of the universe is very bright, thanks to the new revelations from the James Webb Space Telescope and future projects like Artemis and travel to the Moon and Mars.

The James Webb Space Telescope is the most powerful telescope ever built, and it is already providing us with new insights into the universe. For example, the telescope has revealed the deepest and sharpest infrared image of the distant universe ever taken, and it has also captured the first direct image of an exoplanet outside of our solar system.

These discoveries are just the beginning. As the James Webb Space Telescope continues to collect data, we can expect to learn even more about the universe, including the formation and evolution of galaxies, the nature of dark matter and dark energy, and the potential for life beyond Earth.

Future projects like Artemis and travel to the Moon and Mars will also help to expand our understanding of the universe. Artemis is NASA's program to return humans to the Moon and establish a sustainable presence there. Travel to Mars is a longer-term goal, but it is one that is becoming increasingly achievable.

By exploring the Moon and Mars, we can learn more about the history and evolution of our solar system, and we can also search for signs of past or present life. Additionally, these missions will develop new technologies and capabilities that will enable us to explore the universe even further in the future.

Overall, I am very optimistic about the future of our understanding of the universe. The James Webb Space Telescope and future projects like Artemis and travel to the Moon and Mars will help us to make new discoveries and to expand our knowledge of the cosmos in ways that we can only imagine today.

Here is how our understanding of the universe may evolve in the future as a result of these new projects:

We may learn more about the formation and evolution of the early universe.

We may learn more about the nature of dark matter and dark energy.

We may discover new types of planets and galaxies, and even life beyond Earth.

We may develop new technologies that enable us to travel further into space and to explore the universe in new ways.

Looking to the Horizon: The Promise of Tomorrow's Cosmic Explorations

The unveiling of the universe through the lens of the James Webb Space Telescope (JWST) is not an endpoint, but a glowing beacon on our ever-evolving cosmic journey. As we stand at the precipice of unprecedented discoveries, the tapestry of our understanding grows richer, more complex, and yet more wondrous.

JWST, with its deep dives into the early universe, will challenge our existing theories and often, the most profound insights emerge from the questions we hadn't thought to ask. From understanding the birth of stars to potentially uncovering hints of life on distant exoplanets, JWST is bound to reshape our place in the cosmic narrative.

Yet, the cosmos isn't just about distant galaxies and stars; it's also about our neighboring celestial bodies. Projects like Artemis reignite humanity's bond with the Moon. As we prepare to set foot again on our lunar companion, we're not just looking at a mission; we're witnessing the beginning of a new era. The knowledge we gain from the Moon, from its resources to its potential as a gateway for deeper space exploration, can propel us towards our next frontier: Mars.

The Red Planet, with its tantalizing mysteries and history of water, represents a monumental step for humanity. Colonizing Mars, understanding its geology, and seeking signs of ancient life pushes the boundaries of what we believe is possible. It's not just about the science; it's about the human spirit—our innate need to explore, conquer challenges, and expand our horizons.

But it's crucial to remember that our advances in space exploration and understanding aren't just about "out there." Every discovery, every mission, reflects back on us. As we expand outward, we gain perspective on Earth, on our challenges, and on the unity required to meet them.

The future, catalyzed by endeavors like JWST and Artemis, promises a confluence of science, philosophy, and introspection. In unveiling the cosmos, we're also unveiling ourselves—our potential, our aspirations, and our destiny. As we push the boundaries of space, we're also pushing the boundaries of understanding, compassion, and collective growth.

In the dance of stars and planets, in the whisper of galaxies, and in the footprints we'll leave on lunar and Martian soils, lies a message for future generations: We dared, we dreamt, and we discovered. And in doing so, we lit the way for them to dream even bigger.

I am excited to see what the future holds for our understanding of the universe. It is a truly amazing time to be alive.

The next big telescope that will be launched is the Nancy Grace Roman Space Telescope, which is scheduled to launch in 2027. The Roman Space Telescope is a wide-field infrared telescope that will be used to study the dark universe, including dark matter and dark energy. It will also be used to study exoplanets and the early universe.

The Roman Space Telescope is often compared to the Hubble Space Telescope, but it is much more powerful. The Roman Space Telescope has a mirror that is 2.4 meters in diameter, which is slightly larger than the Hubble Space Telescope's mirror. However, the Roman Space Telescope's mirror is also much colder than the Hubble Space Telescope's mirror, which allows it to see fainter objects in more detail.

Another major difference between the Roman Space Telescope and the Hubble Space Telescope is that the Roman Space Telescope is designed to be a survey telescope. This means that it will be used to take images of large areas of the sky, rather than focusing on individual objects. This will allow the Roman Space

Telescope to map the dark universe and to study exoplanets in unprecedented detail.

In addition to the Roman Space Telescope, there are a number of other large telescopes that are currently under development or that are planned for launch in the next decade. These include:

-The Large Synoptic Survey Telescope (LSST), which is scheduled to launch in 2024. The LSST is a wide-field optical telescope that will be used to map the entire sky every few nights.

-The Very Large Telescope (VLT), which is already in operation, but is currently being upgraded to include a new instrument called the Extremely Large Telescope (ELT). The ELT will be the largest optical telescope in the world, and it will be used to study the distant universe and exoplanets.

-The Thirty Meter Telescope (TMT), which is scheduled to launch in 2030. The TMT will be the largest infrared telescope in the world, and it will be used to study the distant universe and exoplanets.

These are just a few of the large telescopes that are currently under development or that are planned for launch in the next decade. These telescopes will revolutionize our understanding of the universe and will help us to answer some of the most fundamental questions about our existence.

Our understanding of the universe is constantly evolving. New discoveries are made all the time, and our theories about how the universe works are constantly being updated.

One of the biggest questions in cosmology today is the fate of the universe. What will happen to it in the long run? Will it continue to expand forever? Or will it eventually reach a point where it collapses back in on itself?

The current evidence suggests that the universe is expanding at an accelerating rate. This means that the distance between galaxies is increasing all the time. If this expansion continues forever, the universe will eventually become so large and diffuse that there will be no more galaxies or stars.

However, there are some theories that suggest that the universe may eventually reach a point where it stops expanding and begins to contract. This would be known as a "big crunch." One theory that suggests this is the Big Bounce theory, which states that the universe has expanded and contracted many times before and will continue to do so forever.

Another possibility is that the universe will eventually reach a state of equilibrium, where the expansion and contraction forces are balanced. This would be known as a "static universe." However, there is no evidence to support this theory at the moment.

So, what does the future hold for our understanding of the universe? It is difficult to say for sure. However, one thing is certain: our understanding will continue to evolve as we make new discoveries.

Here are some specific thoughts on how our understanding of the universe may evolve in the future:

We may learn more about the nature of dark matter and dark energy, which are two mysterious substances that make up the majority of the universe.

We may discover new types of particles and forces that we don't even know about yet.

We may learn more about the history and evolution of the universe, including the nature of the Big Bang.

We may discover new planets and galaxies, and even life beyond Earth.

Tribute to the Guardians of Knowledge

In the boundless realms of the cosmos, where stars weave tales of ancient times and galaxies hum the melodies of creation, a sentinel rises—The James Webb Space Telescope. As we gaze upon its intricate design and celebrate its monumental discoveries, let us take a moment to honor the real stars of this celestial story: the people behind JWST.

To the scientists who dreamt, whose insatiable curiosity became the bedrock of this mission—your vision lit the path.

To the engineers whose genius transformed intricate concepts into tangible, groundbreaking technology—you are the architects of wonder.

To the technicians who meticulously crafted, tested, and refined each component—you ensured the symphony of the cosmos would play without missing a note.

To the managers and administrators, whose unwavering commitment and guidance turned challenges into stepping stones—you were the compass during stormy nights.

To the teams at NASA, ESA, CSA, and every collaborating agency and partner around the world—your unity magnified our potential, proving that when humanity stands together, the sky is not the limit.

To the educators and communicators, who kindled the spark of knowledge in the hearts of the young and old alike—you bridge the expanse between the stars and our souls.

And to every unsung hero whose dedication, hard work, and passion went into this mission—you are a testament to the human spirit's indomitable drive to explore, understand, and transcend.

As JWST unveils the secrets of the universe, let it stand as an enduring tribute to all of you. For in every captured photon and every unveiled mystery, there echoes the collective heartbeat of thousands who dared to dream.

For the sake of knowledge, for the love of discovery, and for the legacy of humanity's relentless pursuit of understanding—thank you.

May the James Webb Space Telescope's journey through the stars forever remind us of the luminous trail of collaboration, perseverance, and wonder that brought it to life.

To all who have dedicated their time to further human understanding,

I salute you.

ABOUT THE AUTHOR

Distinguished astrophysicist, fervent stargazer, and masterful storyteller, Prof. Orion Lumen has dedicated his life to unraveling the mysteries of the cosmos. With a name reminiscent of the stars, it's no surprise that his passion lies in the vast, infinite expanses of the universe. Holding a chair at the prestigious Celestial Institute of Astronomy, Lumen's groundbreaking research has been pivotal in shaping our modern understanding of galaxies, black holes, and nebulae.

But Prof. Lumen's impact isn't confined to the academic realm. His books, a harmonious blend of intricate science and poetic wonder, have captivated millions. Through his eloquent prose, readers voyage to the farthest reaches of space, diving into supernovae, dancing on the rings of Saturn, and chasing comets across the night sky. His best-selling series, "Galactic Whispers," became an instant classic, illuminating young and old minds alike, and sparking a new generation of space enthusiasts.

More than just an astronomer, Prof. Lumen is a beacon in the world of science communication, proving that the wonders of the universe can be both deeply complex and beautifully accessible. His works are not just books but doorways, inviting everyone to gaze upwards and dream beyond the stars.

Manufactured by Amazon.ca
Bolton, ON

38744461R00055